Ederson da Silva Stelato
Nikele Maiara Milani
Gabriele M. Stunges

Stellar chemistry and apocalyptic stars

Ederson da Silva Stelato
Nikele Maiara Milani
Gabriele M. Stunges

Stellar chemistry and apocalyptic stars

Comparative study

ScienciaScripts

Cover image: www.ingimage.com

This book is a translation from the original published under ISBN 978-3-330-76725-6.

Publisher:
Sciencia Scripts
is a trademark of
Dodo Books Indian Ocean Ltd. and OmniScriptum S.R.L publishing group

120 High Road, East Finchley, London, N2 9ED, United Kingdom
Str. Armeneasca 28/1, office 1, Chisinau MD-2012, Republic of Moldova, Europe
Managing Directors: Ieva Konstantinova, Victoria Ursu
info@omniscriptum.com

Printed at: see last page
ISBN: 978-620-8-56670-8

We dedicate this work to our parents for their example, encouragement, love and affection.

ACKNOWLEDGEMENTS

We would especially like to thank Professor Maria Salete Vaceli Quintilio, an astronomer from the University of São Paulo - **SP, for her encouragement and support in** preparing this work, without whom it would not have been complete. **The word master will never do** justice **to dedicated** teachers **who,** without naming them, will have our eternal thanks.

We would also like to thank Unoeste - Universidade do Oeste Paulista de Presidente Prudente - **SP, its teaching staff, management and administration, which provided the opportunity to** carry out this work.

SUMMARY

PREFACE

All the atoms in the Universe, except Hydrogen (H) and Helium (He), come from the processes that take place in stars and their evolution. Stars transmute H and He into other heavier elements through nuclear fusion and their final evolutionary process, supplying and enriching the interstellar medium from which planetary systems originate. However, the same evolutionary cycle that can give rise to a habitable planet can also cause its destruction, along with the possible life forms present, for example in supernova explosions, the evolutionary end of very massive stars.

One star in particular, known as Wolf-Rayet 104 (WR 104), located just 8,000 light years away, is close enough to Earth to generate uncertainties and dire predictions.

Furthermore, when our own Sun reaches the end of its stellar life, it will have fatal consequences for planet Earth, which will eventually have its integrity compromised. When the Sun enters its red giant phase, it will destroy the entire habitable zone of the Solar System, from Mercury to Mars.

In addition, super giant stars, when they "die", can cause mass extinction on the planet, such as WR 104, which could bombard the Earth with intense jets of gamma radiation, again jeopardising the planet's habitability.

It is therefore very important to know and study the evolutionary processes in stellar interiors and then understand how this same chemical evolution in the stars could also jeopardise the habitability of the planet and lead to the end of life on Earth.

CHAPTER 1

INTRODUCTION

The universe is characterised by being made up of matter and energy, with a wide variety of elements, from subatomic particles to exuberant galaxies, originating from the Bing Bang and the evolution of primordial objects. One of the structural components of the universe are the stars, bodies that aggregate a large amount of mass, much larger than planets and capable of generating energy that travels through interstellar space in the form of electromagnetic waves such as light and radio waves, for example.

Stars not only stand out for being energy-producing bodies, but also for the fact that the vast majority of the atoms that make up planets, asteroids, moons and even living beings are left over from dying stars and cosmic dust. In other words, the chemical elements that make up our world almost all come from nuclear furnaces in the interiors of stars and colossal explosions, making stars the protagonists in synthesising chemical elements and enriching the interstellar medium.

The process that takes place inside stars, responsible for producing energy and forming new chemical elements, is nuclear fusion, which is the joining of lighter atoms to form larger and more stable atoms, releasing a large amount of energy and giving rise to a new chemical element. In the Sun, the star closest to Earth, as well as in many others, the process occurs constantly, with hydrogen atoms fusing, forming helium and releasing energy, as shown in the equations below:

$${}_1H^2 + {}_1H^2 \rightarrow {}_1H^3 + {}_1H^1 + \text{energia}$$

$$_1H^2 + _1H^2 \rightarrow _2He^3 + _0n^1 + \text{energia}$$

$$_1H^2 + _1H^3 \rightarrow _2He^4 + _0n^1 + \text{energia}$$

And what maintains the stability of a star? It's the balance between two competing forces: the gravitational force, which tries to make the star contract and collapse, and the radiation pressure, which tends to expand the star due to the flow of energy produced by nuclear fusion.

There are dwarf stars, like the Sun (solar mass $= M_\odot =$ **1.98892*1030** kg), which is a yellow dwarf, and giant and supergiant stars, which can have a mass of between 0.08 and 100 M_o or even over 100$M_\odot$. Dwarf stars are characterised by their long lifespan of around 10 billion years, as they burn their nuclear fuel slowly, while giant and supergiant stars have a short lifespan of around 1 million years.

Dwarf stars produce light chemical elements and when they are at the end of their life, they increase in volume, giving rise to a red giant, and subsequently end up ejecting their outer layers to form planetary nebulae, in the centre of which remains a small, dense stellar remnant known as a white dwarf.

Giant and supergiant stars, on the other hand, produce heavier chemical elements in their cores and, at the end of their lives, the stellar cores collapse because the force of gravity ends up overcoming the radiation pressure from the fusion process and the star loses its equilibrium, ejecting its outer layers in a huge explosion called a nova or supernova, depending on the star's initial mass. In these explosions, even heavier chemical elements are synthesised and the result of this explosion is a neutron star or a black hole, which can emit jets of long-lasting gamma radiation.

The colour of a star is determined by its surface temperature, according to Planck's Law for a black body. The spectra obtained provide its chemical composition, as well as the colour of the gases expelled during the planetary nebula or supernova phase.

It is quite common for young stars to belong to clusters, close to the centre of galaxies, and old, dying stars to be more isolated and on the outskirts of galaxies. They can also form binary systems, in which two stars orbit around a common point, held together by the force of gravity.

As the most abundant atom in the universe, hydrogen is the initial raw material for the production of the other chemical elements inside the stars. Thus, in trying to understand the chemical evolution of stars, we are actually trying to understand the origin and constitution of all the matter that arose after the Big Bang (the most widely accepted theory about the origin of the Universe). It is also from the study of stellar interiors and the synthesis of elements that we seek to understand and reproduce the process of nuclear fusion which, as a means of energy production, surpasses the fission process, since it can produce more energy without resulting in radioactive waste.

In addition, human understanding of stellar evolution, from its formation to its death, is important, because the cataclysmic end of some stellar types, as well as enriching the Universe with heavy atoms, can also cause mass destruction of planets and other stars in their neighbourhood.

In novae and supernovae, the violent endings of massive stars, as seen above, the intense emission of radiation, mainly in the form of gamma ray jets, releases extraordinary energy in intervals of time of the order of just a few seconds. According to a study carried out[1] , a gamma-ray burst in the vicinity of the Earth could be extremely damaging to ocean plankton at depths of up to 75 metres, leading to their destruction. As these microorganisms are responsible for 40 per cent of the photosynthesis produced by the oceans, their destruction could have a serious impact on carbon dioxide levels on Earth.

[1] Penate et al. (2010)

Among the currently known stars that could jeopardise life on Earth is a giant star in the constellation of Sagittarius, 8,000 light-years from Earth, called Wolf Rayet 104 (WR 104). When this star is in its final phase of evolution, becoming a Supernova, it could jeopardise life on Earth in the event of a long gamma-ray burst. Another of them, WR 22, in the constellation of Carina, is one of the brightest and most massive stars known, and its mass has been estimated to be at least 70 times the mass of the Sun and will also generate a Supernova.

Wolf-Rayet stars have masses above $25M_{O}$ and temperatures of around 30,000 to 60,000 K[2] . They are variable stars and have a shroud of dust and gas around them that has been ejected from the star by strong radiation pressure, at a rate of approximately 2 to $10x10^{-5}M_{\odot}$ /year. Around 350 WRs have already been detected in other galaxies.

Another star to consider is the Sun, the star closest to Earth, which, when it becomes a red giant, will evaporate all the Earth's water and calcine its soil, making it impossible for life as we know it to continue. In addition, the cycles of solar activity, which have many harmful effects when at their maximum, must also be taken into account in studies that seek to understand the maintenance of Earth's habitability.

The Sun is currently in the Main Sequence (SP), which is characterised by the fusion of H in its core. It is estimated that the Sun will remain in this phase for 4.5 billion years, after which it will become a Red Giant. WR 104, on the other hand, is a very massive star, around $25M_{Q}$, whose rotational axis is aligned with the Earth. This supernova candidate star is emitting intense and massive stellar winds in our direction.

In the Main Sequence, the star, regardless of its initial mass, fuses hydrogen into helium until the hydrogen runs out and the star starts producing

[2] 1 K = 1 kelvin - (0°C equals 273 K).

new elements such as carbon, nitrogen and oxygen, in the so-called C-N-O cycle, until it obtains iron in its core. After this phase, low-mass stars like the Sun become Red Giants, gradually and slowly losing mass into space until a hot, dense core remains, known as a white dwarf. More massive stars, such as WRs, have a much faster evolution and go supernova, which promotes the formation of even heavier elements and often releases jets of gamma radiation from their poles of rotation.

The study of the chemical evolution of a star answers many questions about the origin of atoms and the formation of planet Earth, the only place in the Universe where we are certain there is life.

In this text we'll deal with two stars: WR 104 and the Sun itself. The processes that can occur inside stars of this class are analysed and compared quantitatively and qualitatively. Data on the stellar wind and the emission of radiation and particles and their possible influence on the Earth's habitability zone are also compiled and analysed.

CHAPTER 2

STELLAR NUCLEOSYNTHESIS

The chemical elements that make up the Periodic Table are produced through nucleosynthesis in stars, supernova explosions or the Big Bang itself.

Figure 1 shows the periodic table with the chemical elements that are synthesised through the Big Bang, supernovae, stars and cosmic rays.

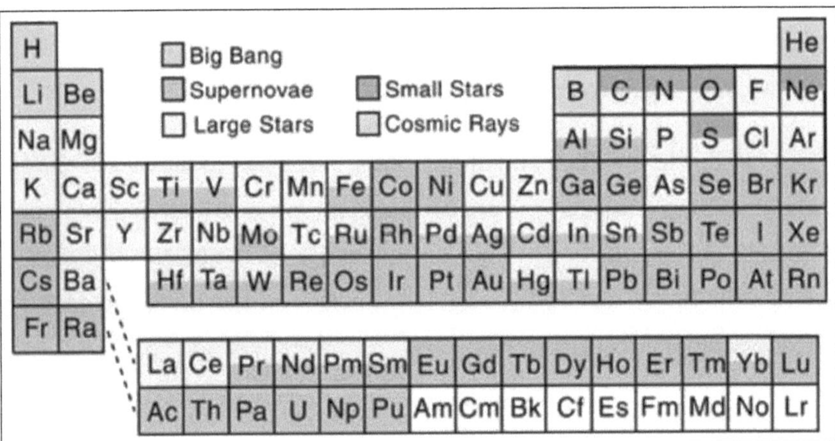

Figure 1 - Nucleosynthesis of the elements (Source: Nerlich, 2015. Credit: North Arizona University)

This figure shows the origin of each chemical element in the periodic table: hydrogen (H) and helium (He) are primordial elements, i.e. they were synthesised during the Big Bang. It can be seen that all stars are basically made up of H and initially produce He. Stars with

Larger stars can continue the process of nucleosynthesis with heavier elements such as copper (Cu), zinc (Zi) and ytterbium (Yb). The elements such as Lithium (Li), Beryllium (Be) and Boron (B) come from cosmic rays; supernovae are the only means of producing all the elements in the periodic table.

As atoms are formed and dispersed into space by stellar winds and supernovae, the interstellar medium is enriched and subsequently the interstellar clouds, the nurseries of stars and planets, incorporate these elements into their formation.

Thus, the first generation of stars, born after the Big Bang, was made up exclusively of H and a little He. As the stellar generations evolved and died, the enrichment of the interstellar medium allowed the heavier elements obtained from nucleosynthesis to disperse.

Figure 2 shows the Hertzprung-Russell (H-R) diagram, devised by Ejnar Hertzprung and Henry Norris Russell in 1910. The H-R diagram illustrates the relationship between a star's luminosity, colour, temperature and absolute magnitude. These relationships characterise a classification of stars and also allow hypotheses and theories about stellar evolution to be tested using observational data and computer models.

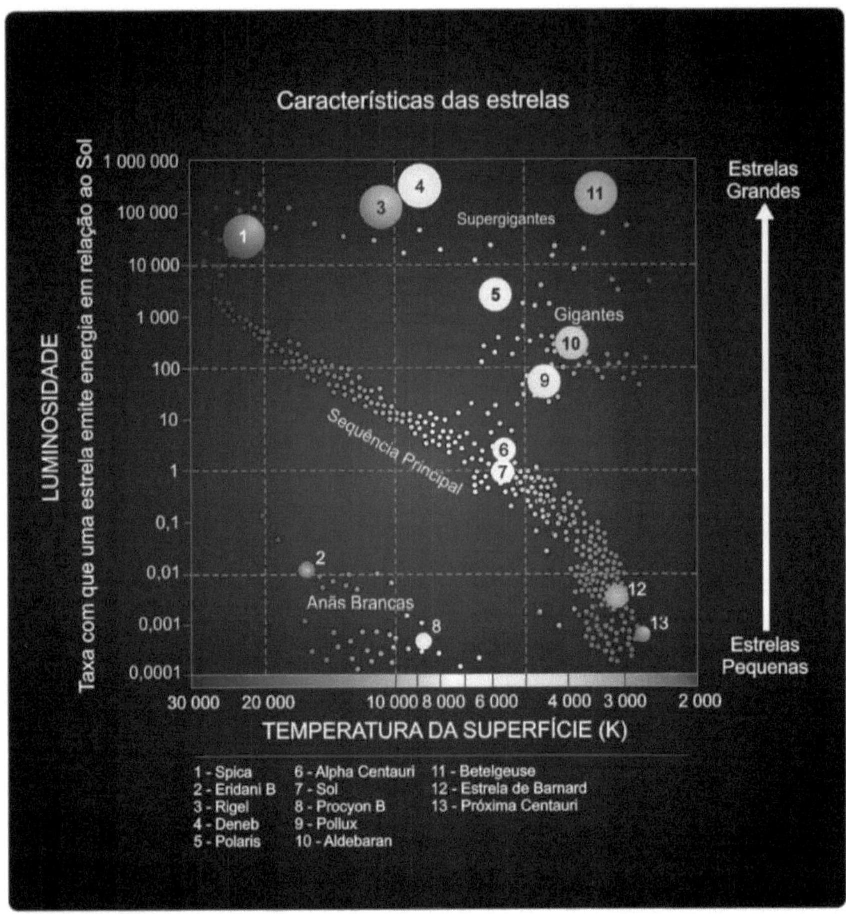

Figure 2 - HR diagram (Source: OLIVEIRA FILHO; SARAIVA 2016)

We can see in Figure 2 that most of the stars are concentrated in a diagonal strip that cuts across the diagram, called the main sequence, which runs from the top left, where the hottest and most luminous stars are located, to the bottom right, where the coolest and least luminous stars are located.

The location of a star on the H-R diagram is determined by its mass, because the greater the mass, the hotter and brighter the star.

All stars begin life on the main sequence by fusing H into He, regardless of their mass. Whether they stay in this phase depends on their mass, as the more massive ones burn their hydrogen more quickly.

Stars approach the final stage when, after the fusion of nuclear hydrogen, they begin to burn the hydrogen in the outer layers and then leave the main sequence.

Figure 3 illustrates the sequence in which stars evolve from beginning to end, depending on their mass.

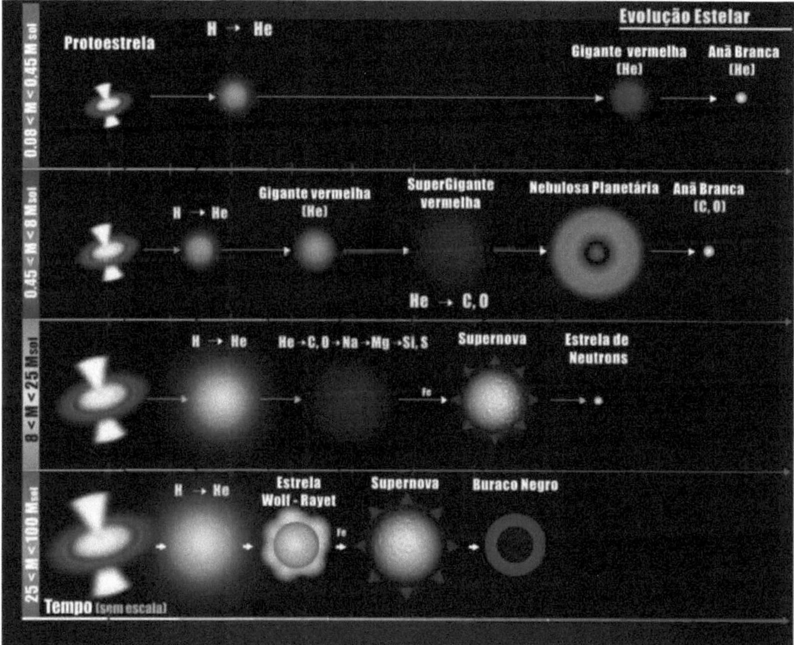

Figure 3 - Stellar evolution, without scale (Source: OLIVEIRA FILHO; SARAIVA, 2016)

The most common fusion process, which transforms H into He, is called the proton-proton (p-p) cycle. Nuclear fusion happens when the kinetic energy of the two nuclei, both positively charged, is enough to overcome the electrostatic repulsion between them. To trigger this process, the star needs to reach temperatures of around 8 million Kelvins.

Equation 1[3] represents the p-p cycle, where H atoms are transformed into He atoms, producing neutrinos (ν_e) and positrons (e^+), and energy in the form of gamma rays (γ).

$$4H \rightarrow He^4 + 2\ e^+ + 2\ \nu_e + \gamma \qquad \text{Equação 1}$$

The amount of energy released in this reaction is more than a million times greater than in a typical chemical reaction, such as burning cooking gas or petrol.

Figure 4 shows this same cycle, but in an illustrative way. The figure shows that four H nuclei come together to form one He nucleus, releasing a large amount of energy.

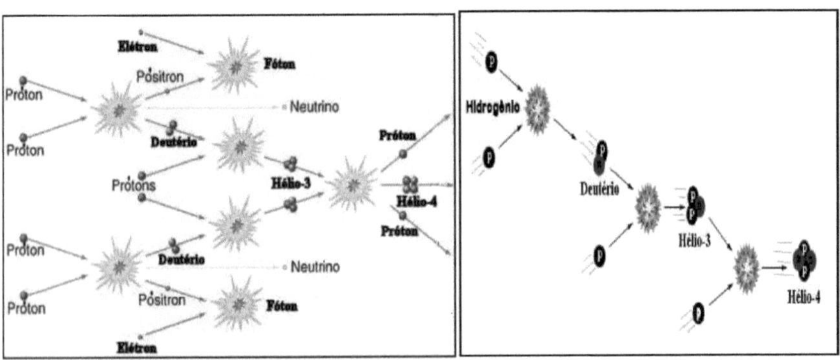

Figure 4 - Proton-proton cycle (Source: Oliveira Filho, 2013)

However, another nuclear fusion process takes place in the most massive stars, called the Carbon Cycle. Figure 5 shows the nuclear reactions that take place in the interiors of stars whose temperatures are higher than the core temperature of the Sun.

[3] According to Caibo 2001.

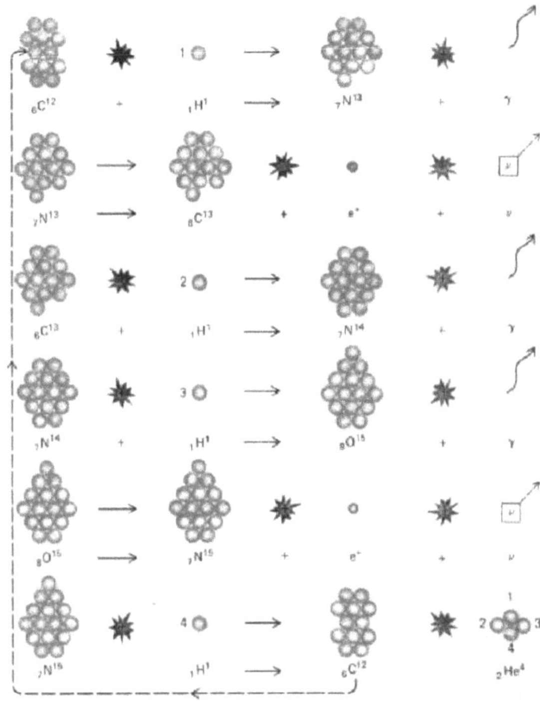

Figure 5 - Carbon Cycle (Source: SANTOS, 2011)

The fusion of carbon begins when it collides with an H nucleus, producing N and energy. At the end of the reaction, after producing O, the final products are He and C nuclei, and the process starts all over again. This cycle, also called the carbon-nitrogen-oxygen cycle (C-N-O), requires temperatures of over 18 million K, which occur in stars with an initial mass of over $1.2M_\odot$. These stars use the elements carbon and nitrogen as catalysts for the process and the reactions are shown in the equations below[4] , where the amount of energy released in each of them (Q) is given, in millions of electron volts (MeV) :[5]

$^{12}C + p \rightarrow {}^{13}N + \gamma$	Q = 1,944 MeV	Equação 2
$^{12}N \rightarrow {}^{13}C + e^+ + \nu_e$ (0,710 MeV)	Q = 1,511 MeV	Equação 3
$^{13}C + p \rightarrow {}^{14}N + \gamma$	Q = 7,550 MeV	Equação 4

[4] Oliveira Filho and Saraiva, 2008, 2016.

[5] 1 electron volt (eV) is the kinetic energy acquired by an electron subjected to a potential difference of 1 volt. It is equivalent to $1.6x10^{(-19)}$ Joules.

$^{14}N + p \rightarrow {}^{15}O + \gamma$	Q = 7,290 MeV	Equação 5
$^{15}O \rightarrow {}^{13}N + e^{+} + \nu_e$ (1,000 MeV)	Q = 1,761 MeV	Equação 6
$^{15}N + p \rightarrow {}^{12}C + {}^{4}He$	Q = 4,965 MeV	Equação 7

The total energy released in this cycle (equations 2 to 7) is 25.02 MeV. For comparison, the energy needed to ionise the H atom is only 13.6 eV. Other reactions can occur, but with a lower probability, as shown in equations 8 to 11:

$^{15}N + p \rightarrow {}^{16}O + \gamma$	(Q = 12,126 MeV)	Equação 8
$^{16}O + p \rightarrow {}^{17}F + \gamma$	(Q = 0,601 MeV)	Equação 9
$^{17}F \rightarrow {}^{17}O + e^{+} + \nu_e$ (0,94 MeV)	(Q = 2,762 MeV)	Equação 10
$^{17}O + p \rightarrow {}^{14}N + {}^{4}He$	(Q = 1,193 MeV)	Equação 11

Equations 2 to 11 occur in stars with initial masses between 0.8 and 10$M_\odot$ which, after consuming nuclear hydrogen, begin the red giant phase and then the red supergiant phase. In these stars, the core is not hot enough to continue the fusion process transforming carbon into heavier elements, and the star passes into the planetary nebula (PN) phase, ejecting its outer layers.

The star then begins to contract due to the gravitational force, as the thermal pressure is no longer sufficient to stop the gravitational collapse, until the temperature stabilises and energy production restarts in the outer layers, which still have hydrogen and helium to fuse. This process causes instability which eventually results in the outer layers of the star being expelled, which forms the NP and the star ending its evolutionary trajectory as a white dwarf.

In more massive stars, above 8 solar masses, the temperatures are enough to fuse carbon into heavier elements and the star passes into the red giant phase after starting the fusion of He in its core. This process takes place without drastic changes and the star burns its fuel in layers, as if it were an onion, as shown in the picture:

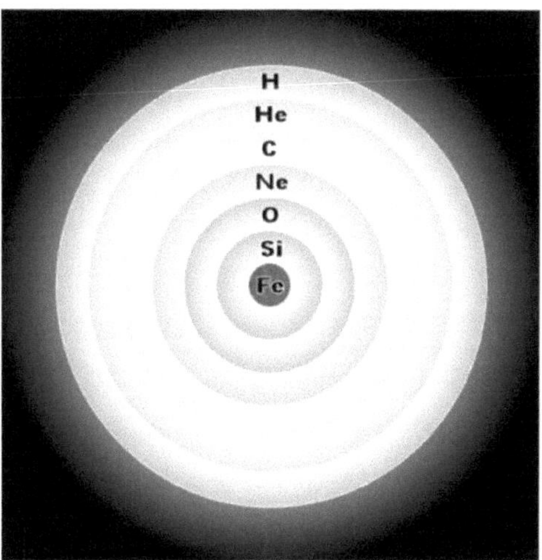

Figure 6 - Layers of a massive star (Source: Noel, 2012)

From this moment on, when nuclear fusion ceases, the stability of the star is lost. As iron forms in the core, the star no longer produces energy and collapses in a supernova explosion, releasing heavier chemical elements into the interstellar medium. The remaining object, which is extremely dense and hot and tiny, is a neutron star or black hole. The nuclear reactions that take place from carbon to iron, when the temperature at the centre of the star is between half and 1 billion kelvins (0.5 - $1x10^9K$), can be described as follows:

$^{12}C + ^{12}C \rightarrow \alpha + ^{20}Ne$ (Q = 4,6168 MeV) Equação 12

$^{20}Ne + \alpha \rightarrow \gamma + ^{16}o$ Equação 13

$^{20}Ne + \alpha \rightarrow \gamma + ^{24}Mg$ Equação 14

$^{24}Mg + \alpha \rightarrow \gamma + ^{28}Si$ Equação 15

$^{12}C + ^{12}C \rightarrow p + ^{23}Na$ (Q = 2,2398 MeV) Equação 16

$^{23}Na + p \rightarrow \alpha + ^{20}Ne$ Equação 17

$^{23}Na + p \rightarrow \gamma + ^{24}Mg$ Equação 18

$^{12}C + ^{12}C \rightarrow n + ^{23}Mg$ (Q = -2,5993 MeV) Equação 19

Or the cycle can occur differently but with less chance, according to the equations below:

$^{12}C + ^{12}C \rightarrow ^{24}Mg + \gamma$ (Q = 13,9313 MeV) Equação 20

$^{12}C + ^{12}C \rightarrow ^{16}O + 2\alpha$ (Q = -0,1132 MeV) Equação 21

$^{12}C + ^{12}C \rightarrow ^{16}O + ^{8}Be$ (Q = -0,2080 MeV) Equação 22

When stellar temperatures reach between one and two billion of kelvins (1 - 2x109 K), the following reactions occur:

$^{16}O + ^{16}O \rightarrow ^{32*}S \rightarrow \gamma + ^{32}S$ (Q = 16,5410 MeV) Equação 23

$\rightarrow \alpha + ^{28}Si$ (Q = 9,5928 MeV) Equação 24

$\rightarrow p + ^{31}P$ (Q = 7,6770 MeV) Equação 25

$\rightarrow n + ^{31}S \rightarrow ^{31}P + e^{+} + \nu_e$ (Q = 1,4531 MeV) Equação 26

$\rightarrow 2\,p + ^{30}Si$ (Q = 0,3795 MeV) Equação 27

For temperatures between 3.4 and 3.7x10^9 K, we will have:

$^{12}C + ^{16}O \rightarrow \gamma + ^{28}S$ (Q = 16,7544 MeV) Equação 28

$\rightarrow p + ^{27}Al$ (Q = 5,1691 MeV) Equação 29

$\rightarrow \alpha + ^{24}Mg$ (Q = 6,7697 MeV) Equação 30

$\rightarrow n + ^{27}Mg$ (Q = -0,4230 MeV) Equação 31

When the temperature reaches values greater than 5 billion kelvins (T > 5x109 K), the process ends with the synthesis of the element iron (Fe):

$^{28}Si(\alpha, \gamma)\ ^{32}S(\alpha, \gamma)\ ^{36}A(\alpha, \gamma)\ ^{40}Ca(\alpha, \gamma)\ ^{44}Ti(\alpha, \gamma)\ ^{48}Cr(\alpha, \gamma)\ ^{52}Fe(\alpha, \gamma)\ ^{56}Ni$

$^{56}Ni + e^{-} \rightarrow \nu_e + ^{56*}Co$ e $^{56}Ni + e^{+} \rightarrow \nu_e + ^{56*}Co$ Equação 32

$^{56*}Co \rightarrow ^{56}Co + \gamma$ Equação 33

$^{56*}Co + e^{-} \rightarrow ^{56*}Fe + \nu_e$e $^{56*}Co \rightarrow ^{56*}Fe + e^{+}\nu_e$ Equação 34

$^{56*}Fe \rightarrow ^{56}Fe + \gamma$ Equação 35

Table 1 shows the amount of energy released in the main nuclear reaction cycles shown above, where a indicates a He nucleus (a particle with 2 protons and 2 neutrons). Remember that these energies are only due to a molten nucleus! The rate at which a star

Table 1 - Amount of energy released in nuclear reactions

Process	QNA / **A** (MeV/nucleus)
4 H→ ^{4}He	5 a 7
3a→ ^{12}C	0,606
4a -o ^{16}O	0,902
2 ^{12}C→ ^{24}Mg	0,52
2 ^{20}Ne→ ^{16}O + ^{24}Mg	0,11
2 ^{16}O→ ^{32}S	0,52
2 ^{28}Si→ ^{56}Ni	0 a 0,31

(Source: Oliveira Filho and Saraiva 2004)

Astronomers Kepler de Souza Oliveira Filho and Maria de Fátima Oliveira Saraiva (2016) compare the energies in some processes in Table 2:

Table 2: Reactions that release energy

	Chemistry	Fission	Fusion
Reaction examples	C + O2→ CO2	n + u^{235}→ Ba^{143} + Kr^{91} + 2 n	H^{2} + H^{3}→ He^{4} + n
Typical fuel	Coal	UO2 (3% U^{235} + 97% U^{238})	Deuterium and Lithium
Temperature for reaction (°C)	873	1273	1 000 000
Energy released per kg of fuel (J/kg)	3,3 x 10^{7}	2,1 x 10^{12}	3,4 x 10^{14}

(Source: Oliveira Filho and Saraiva 2016)

The nucleosynthesis of chemical elements heavier than iron (Fe^{56}) can occur through three types of process: neutron capture (s-process and -r-process) and proton capture (p-process).

Neutron capture begins with the collisions and fusions of neutrons, resulting in nuclei with many neutrons and greater mass, with equal atomic number, forming an isotope. The isotope formed can be stable or unstable. If it is unstable, it can cause spontaneous radioactive decay, called beta decay, where an electron and an anti-neutrino are emitted and, in the nucleus, a neutron becomes a proton. By changing the number of protons, a new element is obtained.

If neutrons are captured by the s-process (or slow process), the unstable atomic nucleus formed undergoes beta-decay before more neutrons are captured, generating an atom with one proton more or less. This process is found in the final stages of stars like the Sun (red giants), because when it turns into a white dwarf, neutrons are produced (mainly from the decay of carbon and neon).

The s-process is capable of producing heavier elements from iron (26 protons), from copper (Cu) with 29 protons to thallium (Tl) with 81 protons.

In the case of neutron capture by the r-process (or fast process), during the collapse of a star with an iron core (supernova), many neutrons are released and with this, the unstable core formed will have more time to capture more neutrons before undergoing beta decay.

Thus, in the r-process, elements are formed that could not be generated by the s-process, ranging from lead (Pb) with 82 protons to plutonium (Pu) with 94 protons.

The other process that takes place is the capture of protons (p-process), which occurs when the protons are hot and contain enough energy to cross the electrostatic barrier that is formed due to the repulsion of the proton when it approaches the nucleus to establish fusion, where new heavier elements are formed. This happens mainly in binary systems, as one of the stars attracts the hydrogen gas from the other, generating protons.

The evolution of normal stars, as seen above, produces the nucleosynthesis of elements up to iron, which is the most stable element in the periodic table. The remaining elements are generated mainly in supernovae and their remnants, which will be discussed in the next chapter.

CHAPTER 3

APOCALYPTIC STARS

3.1 Supernovae

A supernova can reach a luminosity of 100 billion times the solar luminosity. This final event of supermassive stars is sudden, and its brightness can surpass that of the galaxy hosting the star and can last up to a few weeks.

This catastrophic event occurs after the formation of an iron core, which undergoes a violent contraction in just a few seconds, under the effect of gravity that no longer has the radiation pressure to counterbalance it, as the iron no longer produces energy via nuclear fusion. The outer layers of the star collapse onto the core, drastically increasing its temperature, causing the stellar material to be expelled very quickly and violently, reaching speeds of thousands of kilometres per second in a supernova-like explosion.

Supernovae are classified into type I and type II according to their particularities. These types are further divided into subtypes. Thus, SN Ia is the explosion of a white dwarf and Ib, Ic and type II are explosions of giant stars ($M > 9M_0$) and occur in the arms of spiral galaxies.

Type I SN has almost no hydrogen and its luminosity drops rapidly when the explosion occurs, while type II SN, on the other hand, has an abundance of hydrogen and its luminosity decreases very slowly.

A curious fact is that SN type I occurs on average every 100 years and SN type II every 30 years, considering spiral galaxies.

The type Ia supernova is quite peculiar, as it is the revival of a dying star. A white dwarf considered "dead" can "come back to life" when it is part of a binary system. If the white dwarf is associated with a main sequence star or a giant, it can transfer the chemical elements H and He from its companion to itself. As a result, it becomes hotter and denser; when it reaches a temperature of 10 million kelvins (10^7 K), the hydrogen is burnt again, but quickly and violently, increasing its luminosity and the white dwarf shines again, in other words, it comes back to life. This process is known as Nova. The white dwarf adds more material and at some point equilibrium cannot be maintained because the mass has exceeded the stability limit for white dwarfs (1.4 M_Q) and collapse occurs. This increases the temperature and causes helium to fuse into carbon in the entire zone of the star simultaneously, exploding it as a supernova.

The final stage of SN type I produces the supernova remnants, clouds of ionised gas that surround the white dwarf, and type II produces a tiny and very dense stellar nucleus, called a neutron star. The following figures are images of some supernovae.

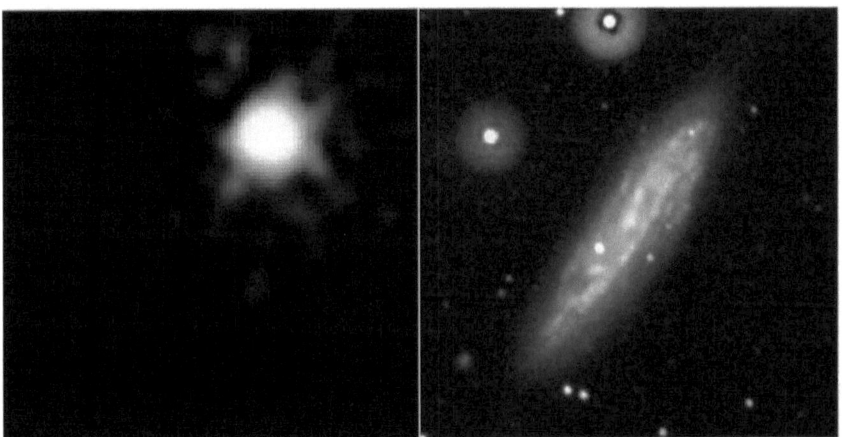

Figura 7 - Type Ib supernova - X-ray (left) and visible (right) photo of SN 2008D, located in the galaxy NGC 2770.(*NASA* **image**. Source: SÉRVULO, 2014)

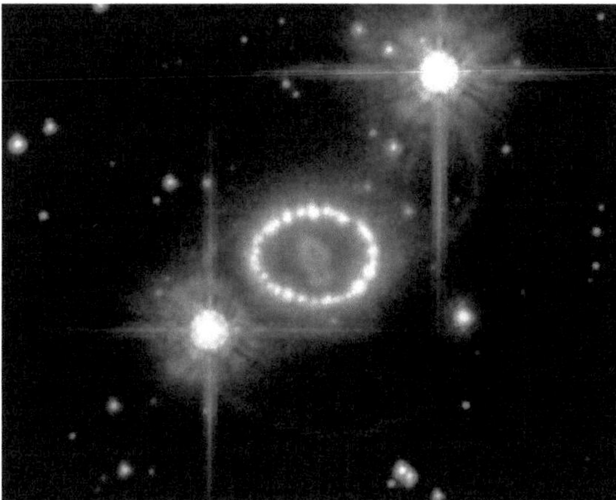

Figura 8 - Type II supernova - The expanding remnant system of SN 1987, a Type II-P supernova, located in the Large Magellanic Cloud. Credit: *NASA* (Source: SERVULO, 2014)

3.2 Neutron stars

Neutron stars are the end products of the evolutionary process of stars considered to be giants (with a solar mass of between 8 and 25). They have a radius of around 10 kilometres, which results in a density of around 10^{15} g/cm^3. At the end of their lives, the thermonuclear reactions that take place in their core cause them to evolve to the red giant stage, a stage in which the star is capable of producing iron. Figure 9 shows an artistic representation of a neutron star.

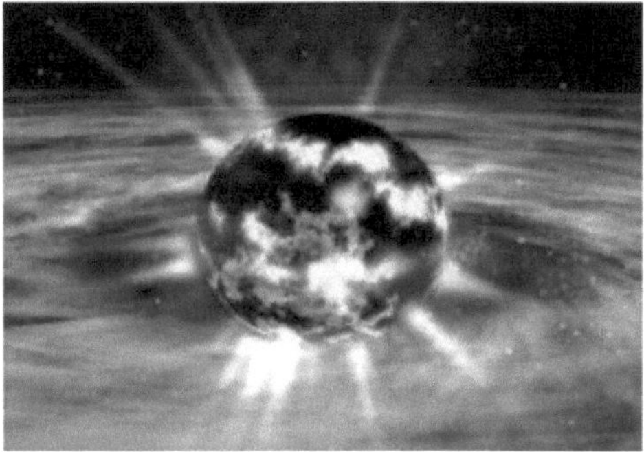

Figure 9 - Artistic representation of a neutron star. Credit: NASA/Dana Barry (Source: NASA, 2007)

At this stage, the nuclear fusion process is still active in the outer layers of Si and O, among others. Fusion in these outer layers increases the mass of the nucleus and gravity compresses it to the point of creating electrons with speeds of the order of the speed of light. At this stage, the relativistic electrons reach the point where they can withstand the increased mass of the nucleus. From this stage onwards, the pressure of the electron gas is not enough to sustain/counterbalance the pressure of the gravitational force on the nucleus and a violent implosion begins, reaching very high temperatures, of the order of MeV (~10^{11} K). Energetic neutrinos with nuclei densities of 10^{12} g/cm^3 are confined inside the nucleus. These neutrinos are carried along with the material falling on the core during its heating, which reaches tens of Mev.

When the star can no longer sustain the material beyond its core, a decompression wave occurs, causing this material to be ejected outwards at high speeds and the star explodes into a supernova.

The material ejected in the explosion, called the supernova remnant, expands at a typical speed of 10,000 km/s. The energy released in a supernova collapse generates a luminosity equivalent to a galaxy with 200 billion stars and causes the stellar remnant to rotate on itself with very high angular velocities, making the star appear to pulsate. Such neutron stars are called pulsars.

Pulsars were discovered in 1967 by astronomy student Jocelyn Bell and her thesis supervisor Tony Hewish, who detected them using radio waves. The radio pulses that were recorded were strangely constant, so much so that some astronomers wondered if they were picking up signals from extraterrestrial civilisations. The correct explanation was even stranger: what was emitting the signals was a neutron star that was spinning very fast and had the mass of a sun in a sphere about 20 kilometres in diameter. Pulsars can rotate thousands of times every hour, emitting radio pulses through their poles in the style of a high-speed beacon.

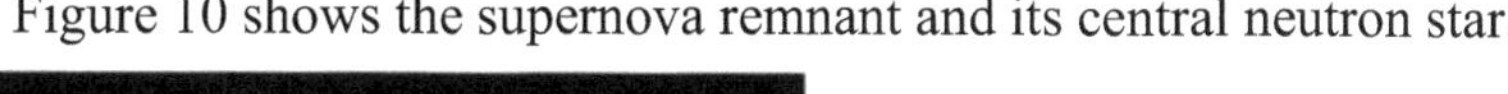
Figure 10 shows the supernova remnant and its central neutron star.

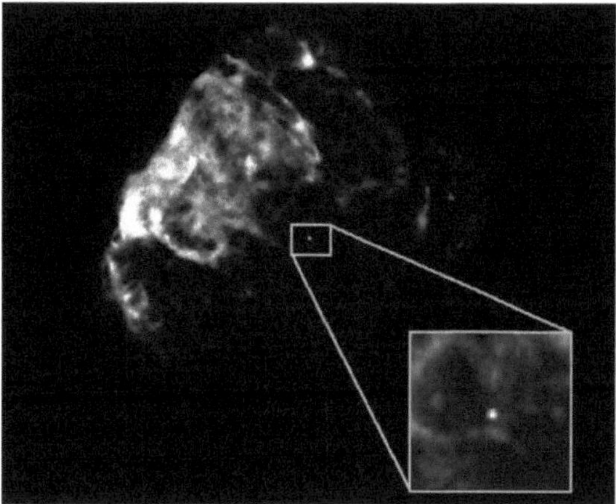
Figure 10 - Supernova remnant and neutron star (Source: NASA, 1998)

Pulsars can remain active for around 10 million years or more. Over time, the pulsar's rotation slows down and its magnetic field decreases, and eventually they become invisible, joining the dark matter (which doesn't shine) present throughout the Universe.

The pulsar present in the Crab Nebula (Figure 11) is a neutron star with a very intense magnetic field and a rotation rate of 30 times per second. The X-ray image gives the first clear view of the boundaries of the nebula and the pulsar. The pulsar's energy accelerates charged particles, producing jets of X-rays directed in the polar direction and an intense wind in the equatorial direction. With more mass than the Sun and the density of an atomic nucleus, this pulsar is the collapsed nucleus that resulted from the explosion of a supernova that was witnessed in the year 1054. This image taken by the Chandra telescope covers just under 9 light-years; the estimated distance of the crab nebula from Earth is 6,000 light-years.

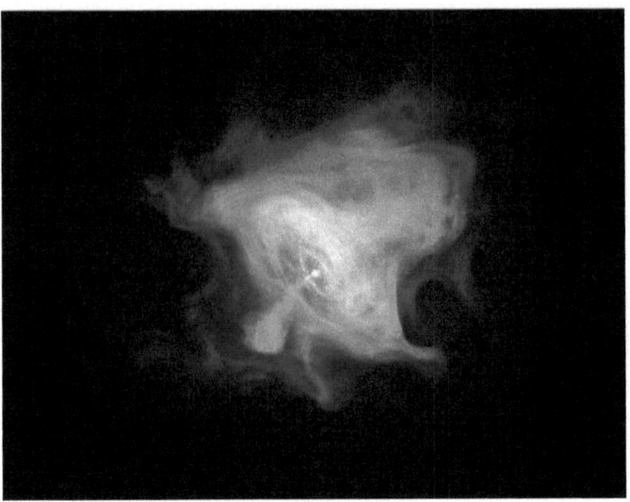

Figure 11- Pulsar and Crab Nebula (Source: NASA, 2008)

In a broader perspective, neutron stars and heavy ion collisions give us access to the phase diagram of nuclear matter at extreme temperatures and densities, which is essential for understanding the first moments of our universe as well as various other astrophysical phenomena.

3.3 Black Holes

Black holes are cosmic objects characterised by an immense gravitational force, which causes a great deformity in the fabric of space-time, pulling any form of matter into its interior

or energy around it. Not even light with its speed of 300,000 km/s can overcome the escape velocity of a black hole, which is why it is invisible to the optical spectrum. Figure 12 shows the intensity of the gravitational field of a black hole compared to other types of objects; we can see the deformity of space-time.

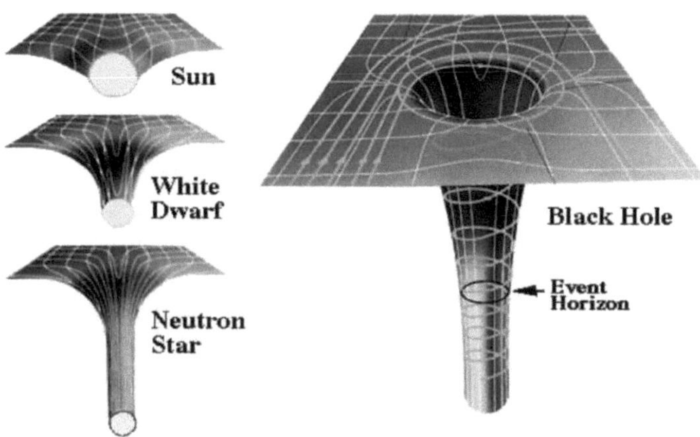

Figure 12 - Illustration showing the deformation of the fabric of space-time by large cosmic objects (Source: SERVULO, 2015)

Black holes originate after supernovae of stars whose masses are between 25 and 100 solar masses, when the core, already made up of iron, is unable to maintain the process of nuclear fusion and therefore balance the force of gravity. The violent collapse of the star results in this gravitational phenomenon. Any matter or form of energy that reaches the event horizon (the region around it) of a black hole is sucked in by it.

Theoretically, there are supermassive black holes at the centre of all galaxies, necessarily resulting from the merger of smaller stellar black holes, since hypergiant stars inhabited the universe even before galaxies were formed. These supermassive black holes have an estimated mass of hundreds to billions of solar masses and have colossal gravitational power. The table below shows the probable galaxies estimated to have a supermassive black hole at their centre.

Table 3 - Galaxies estimated to have a supermassive black hole at their centres.

Galaxy	Black Hole Mass (Sun=1)
IE1740.9-2942	100 hundreds
Sgr A*	2 million
Messier 32	3 million
Centarus A	< 14 million
Messier 31	30 million
Messier 106	40 million

NGC 3379	50 million
NGC 3377	100 million
Messier 84	300 million
NGC 4486B	500 million
NGC 4594	1 billion
NGC 4261	1 billion
NGC 3115	2 billion
Messier 87	3 billion
Cygnus A	5 billion
NGC 4151	Unknown
Messier 51	Unknown

Source: CASAS, 1999.

As black holes are opaque to the visible, they are not observed in images captured by terrestrial and space telescopes, and their identification occurs indirectly, i.e. by observing the orbital behaviour of stars around black hole candidates or by capturing X-ray waves, since as matter is sucked in by a black hole, friction generates the emission of energetic radiation. It is a paradox that, at the same time as the black hole sucks in matter and energy, it releases energy as a result. This happens especially in the case of binary systems, where a star orbits a black hole that gradually consumes the star's plasma, as in the case of the Cygnus X-1 black hole and its blue super giant companion star O9.7Iab HD226868. Table 4 shows the probable stars that have a black hole as a companion.

Table 4 - Stars that have a black hole as a companion in a
a binary system.

Star	Black Hole Mass (Sun=1)
A0620-00	3 - 4
Cygnus X-1 (HDE 226868)	4 - 8
Sco X-1	3 - 10
GS 2000+25	3 - 10
GX 339-4	3 - 10
V 404 Cygni	8 - 12
Nova Muscae 1991	3 -10
Nova Ophiuchi 1977	6 - 7

Source: CASAS, 1999

Figure 13 shows an image of galaxy IC 342 taken by the NuStar high-energy satellite, which emits X-rays typical of black holes.

Figure 13 - X-ray emitted by a black hole in the centre of the galaxy (Source: OLIVEIRA FILHO; SARAIVA, 2016. Credit: NASA)

Figure 14 shows images taken in the infrared region by the European Southern Observatory telescope; the arrow shows a star that has passed for 17 light hours through the black hole in the centre of the Milky Way.

Figure 14 - Star orbiting near the Milky Way's supermassive black hole (Source: OLIVEIRA FILHO; SARAIVA, 2016. Credit: NASA)

In 2013, NASA released an image of a cosmic object called W49B, composed of stellar remnants after a supernova, which astronomers estimate to be the birth of a black hole, as can be seen in the figure below.

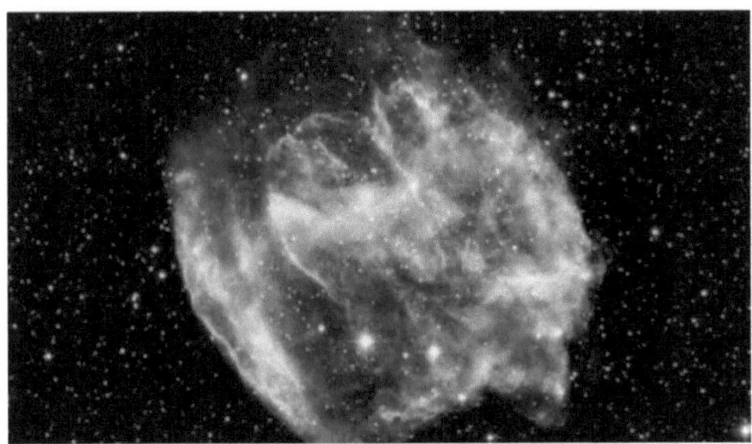

Figure 15 - Birth of a black hole (Source: NASA, 2013)

CHAPTER 4

THE SUN

The central star of our system, which provides the energy and heat that sustains life on Earth, is simply called the Sun.

The Sun is 150 million kilometres from Earth and emits radiation across the entire magnetic spectrum, concentrating its maximum intensity in the visible, ultraviolet and infrared regions. Its mass is M_o = 1.9891x1030 kg, contained within a radius of 6.96×10^6m (100 times the radius of the Earth). These dimensions give the Sun an average density 40% greater than that of water.

The Sun's composition is mostly hydrogen, a little helium and traces of other elements (see Table 5). These gases are heated to high temperatures, forming the ionised solar plasma. On the Sun's surface, the temperature is 5,800 K, but in its centre, temperatures can reach 15,000 K. It is in the centre of the Sun that the thermo-nuclear reactions that transmute hydrogen into helium take place, generating large amounts of energy in the form of photons and thermal movements.

Table 5 - Chemical composition of the sun

Element	Percentage
Hydrogen	92,1 %
Helium	7,8 %
Oxygen	0,061 %
Carbon	0,039 %
Nitrogen	0,0084 %

Source: SILVA, 2006

The Sun can be divided into two regions, the interior and the atmosphere. The interior is made up of: the core, the radiative zone and the convective zone. Its atmosphere (the transition region between the interior and the interstellar medium) is divided into the photosphere, chromosphere and corona. These regions can be seen in Figure 16.

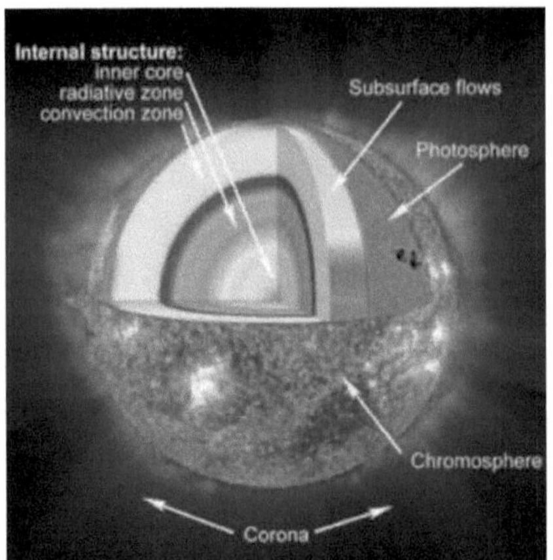

Figure 16 - Layers of the Sun (Source: NASA, (2007)

The nucleus is where nuclear fusion and nucleosynthesis of the elements take place. The radiation produced is directed into the Sun's atmosphere via the radiative and convective zones, where it is then sent out into interplanetary space. Therefore, we will only describe the Sun's external structures, as the observational information comes from these parts.

The photosphere is a narrow layer approximately 500 kilometres thick with a temperature of 5800K and a lower density than the inner layers (~5 x 10^{15} cm^{-3}). The light we see from the Sun comes from the photosphere, as its lower density reduces its opacity.
allows radiation to propagate. The photosphere has a heterogeneous brightness: direct photos show bright granules surrounded by dark spots, which are called solar granulation, formed by masses of hot gas that increase in size and transport energy by convection that will be dissipated in the photosphere and with cooling the gases descend through the dark cells.

Another factor that makes the photosphere a heterogeneous brightness layer is temperature variation. At the edges of the Sun there is a decrease in brightness defined as limbic obscuration. In white light, brightness is maximum in the centre of the solar disc and drops by 20% at the edges. From the base of the photosphere the temperature decreases, increasing again when it reaches the chromosphere.

The spectrum of white light in the photosphere is not continuous, but has gaps in the form of well-defined dark lines. The weakest lines are produced in the inner regions of the photosphere, while the strongest are produced in the outer regions. These lines were first identified by Fraunhofer in 1814, who used capital letters to denote stronger lines and lower case letters for weaker lines. The most studied today are the D lines of the sodium doublet, the H and K lines of Ca II and the B lines of magnesium. Figure 17 shows the historic spectrum obtained by Von Fraunhofer.

Figure 17 - The von Fraunhofer spectrum (Source: LARANJEIRAS, 2010)

Fraunhofer discovered that the black lines were superimpositions of several lines corresponding to the absorption spectra of the chemical elements present in the sun. Looking at Figure 17 and 18, it can be seen that the emission spectra of sodium and helium are very close and correspond to the black lines in the solar spectrum.

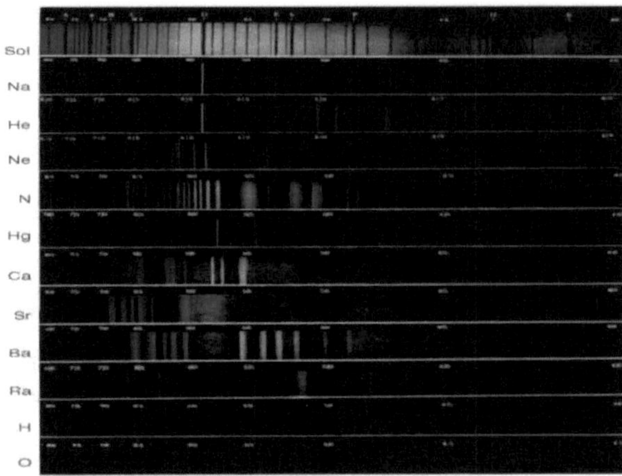

Figure 18 - Solar spectrum and spectrum of various elements obtained in the laboratory (Source: FIGUEIRAS, 1996)

In the photosphere we also find sunspots (Fig.19), which are regions where the magnetic field is more twisted and complex, inhibiting convective transport, making them much colder than the atmosphere. It was by observing that the spots varied their position that Galileo discovered that the sun rotated. Sunspots are formed because the sun has millions of poles which mean that its plasma doesn't rotate evenly and is only held there by its gravity. We know this because the magnetic field lines move the plasma in the form of coronal arcs that rise in the solar atmosphere, moulding the magnetic fields. Sometimes the magnetic fields mould the plasma into helical shapes called flux tubes, which are twisted magnetic field lines containing a lot of free magnetic energy. When they bend, this energy increases, but over time this energy escapes and launches this mass into space. In regions where the magnetic field is more twisted and complex, the rising heat is blocked and causes the temperature to drop by up to 500 degrees, forming sunspots. Sunspots only appear dark because they are close to very bright places, because if they were compared to the full moon, their brightness would be 10 times greater.

These spots form and destroy themselves in a matter of months, during which time they grow until they fragment and disappear. When sunspots release their magnetic energy, the most dramatic explosions occur on the solar surface, which are very bright due to the high temperatures. They appear in groups according to the bipolarity of the magnetic field.

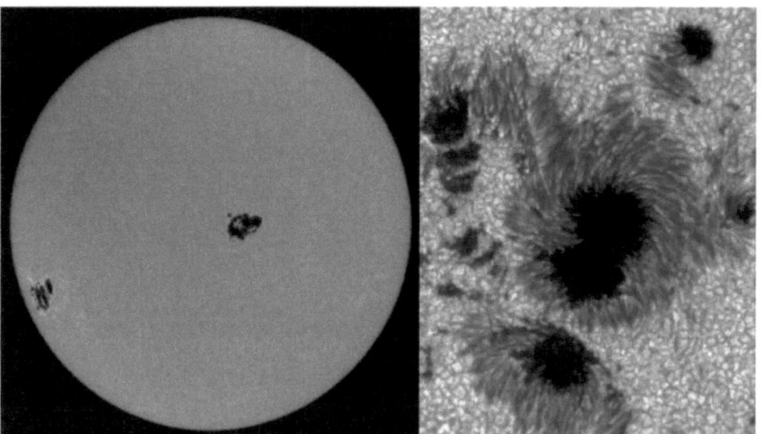

Figure 19 - Sunspots (Source: BERGMAN, 2016. Credits: **SOHO (NASA & ESA) and the Royal Swedish Academy of Sciences).**

The chromosphere has a much lower density than the photosphere and varies from 10^9 cm$^{(-3)}$ on the outside to 10^{12}cm$^{(-3)}$ at its base. Its temperature ranges from 6,000°C to 20,000°C. It has a reddish colour that is only visible when using special equipment or during a solar eclipse. It is a very hot area and it is speculated that this heat is not due to the photons produced in the nucleus but to the magnetic field formed in the photosphere which is transported by electric currents to the chromosphere where its energy is left behind.

Spicules are faint jets of bright gas between 500 and 1500 km in size that reach a height of 1000 km and appear dark because they are colder. They last less than 15 minutes, but are important for the balance of the solar mass, the corona and the solar winds and only occur in regions of increased magnetic field strength.

Between the chromosphere and the corona there is a transition zone where the temperatures rise rapidly from 10^4 to 10^6K. This region is very hot because it is full of convection cells that bring very hot material from the interior of the sun, which rises and then cools down, producing light and returning to the centre. It's a very violent and noisy process, and it's this sound energy combined with the sun's magnetic energy that heats up the corona.

The corona is the outermost part of the sun, with a lower density and higher temperatures. It is made up of plasma, protons and free electrons and can only be seen without specific equipment when there is a total eclipse of the sun. The temperature in the corona can reach 1,000,000°C, and its shape can vary due to the activity of sunspots. In addition, the corona occupies a region more than 1 million kilometres from the sun and is 200 times hotter than the visible surface of the star.

Solar winds are caused by coronal gases that have not been trapped by the sun's gravitational action. The solar wind is made up of protons and electrons in equal proportions, so it is electrically neutral, but its thermal conductivity is very high, making its heat very intense even far from the sun. Solar storms happen when sunspots release their magnetic energy, causing huge eruptions on the sun's surface. When this magnetic energy hits the Earth, it can wreak havoc on electronic devices. Nowadays, when technology is based on electronics, these storms cause a lot of damage.

According to Oliveira Filho and Saraiva (2012), some of the consequences of solar storms are:

> • Damage to satellites, also caused by the increased friction caused by the expansion of the atmosphere,
>
> • Errors in the positioning of ships and aeroplanes of several kilometres, both by the GPS (Global Positioning System) system and by the Loran and Omega systems (8 transmitters distributed around the Earth), due to instabilities in the plasma of the ionosphere terrestrial, causing flickering in the amplitude and phase of the signal and reducing the number of satellites available from 8 to 10 to up to 4. In general, these instabilities last less than 10 minutes, but there have been cases in which the system has been off the air for up to 13 hours,

- Damage to power grids, inducing voltages of thousands of volts and burning out transformers.
- Damage to the metal pipework of gas pipelines, as the induced currents drastically increase corrosion,
- They also increase the incidence of ionising radiation on people, especially on high-altitude flights, such as supersonic and astronautical flights.

As it is a dwarf star, the sun will not explode in a supernova, but it will reach the Red Giant phase, in which it will continue to burn hydrogen to generate energy, fusing them and transforming them into helium and energy. In its final stages, the core becomes poorer in hydrogen and richer in helium. From then on, the core begins to contract, which increases its pressure and temperature, allowing the rate of hydrogen fusion to accelerate, thus increasing the star's luminosity. When the sun's core becomes a mass of helium, its temperature will not be sufficient for the fusion of this element to occur, forcing the core to contract until it reaches approximately the size of the Earth and the temperature rises to 100 million degrees, high enough to fuse helium nuclei into carbon and oxygen.

The beginning of carbon synthesis signals the beginning of the end for small stars like the Sun. At this stage, when the star is dying, the balance between gravity pulling inwards and pressure pushing outwards ends. Gravity begins to contract the core and the external pressure pushes the gases outwards, which expand to form an immense ball of gas called a red giant. The red colour is due to its cooling, as it will expand so much that it will start to cool down, but its luminosity will be very great due to its immense size, reaching

swallow the Earth and nearby planets. Meanwhile, gravity will compress the core to a millionth of its original size and the sun will become a ball of oxygen and carbon called a white dwarf, which is a body in which the electrons are so compressed in the core that a spoonful of this material would weigh several tonnes. The gas that was around it will expand and disperse into the universe, forming a Planetary Nebula.

Figure 20 shows the process of the sun's evolution.

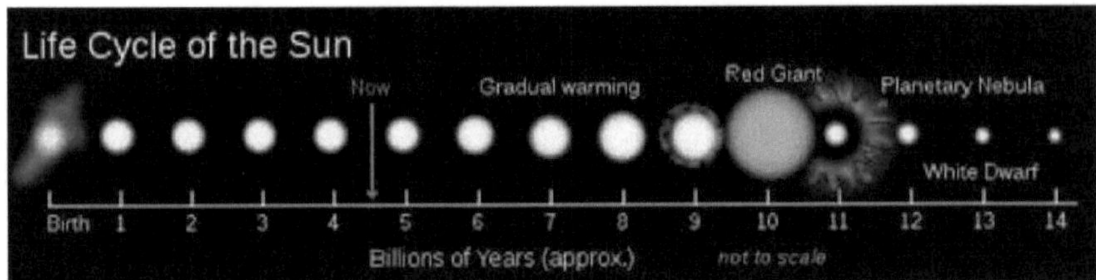

Figure 20 - Illustration of the Sun's evolution. The sizes are not to scale

(Source: CHAVES, 2011)

CHAPTER 5

WOLF-RAYET STAR 104

The star Wolf-Rayet 104 (WR 014) is a massive star located in the constellation Sagittarius, 8,000 light-years from planet Earth. This star is characterised by having a radius 3 times greater than the solar radius and a mass 25 times that of the Sun. The star was discovered in 1998 using the Keck I infrared telescope (Figure 21 taken by P. Tuthill&J. Monnier at the Keck I Observatory in Australia).

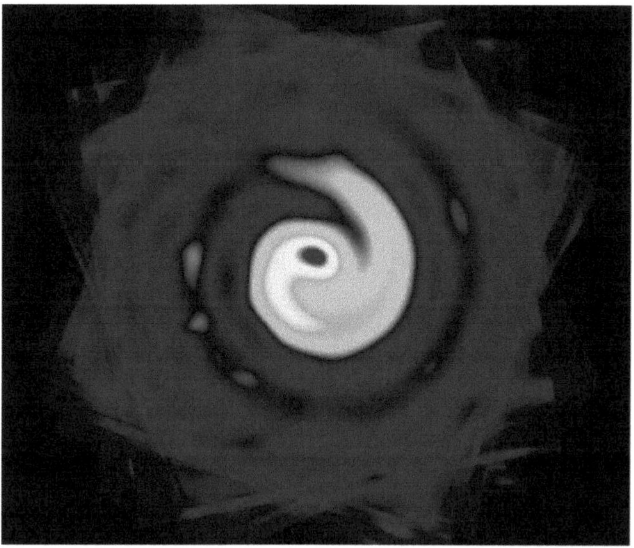

Figure 21 - Wolf-Rayet 104 star (Source: NASA, 2013)

This unusual spiral pattern is created by winds and dust that are expelled by the binary system. One of the components is the Wolf-Rayet star, in its final stage of evolution before going supernova, which will occur over the next million years, and the other is an OB-type giant star.

From the image taken, we can see the presence of a spiral of stellar plasma around the star with an extension of 160 astronomical units (1 A.U. is equal to the distance from the Earth to the Sun). This plasma spiral is the result of the WR star losing mass. In addition, because it is a binary system, the companion star (OB) also releases a certain amount of plasma which, when it collides with Wolf-Rayet's plasma, forms a spiral wind that accompanies the orbital movement of both stars.

WR 104 will undergo a supernova-type explosion, but the time for such an explosion has not yet been determined due to the star's distance from planet Earth, as there is a lot of dust between the star and Earth, making it difficult to determine the star's real age. However, the most important thing is to know what type of supernova the star will generate: if it's an ordinary supernova, where the star's core and remaining material turn into a neutron star (pulsars or magnetars), the explosion won't affect the Earth. At this distance of 8,000 light-years, only a large, sudden glow will be detected in the night sky.

However, if the star suffers a supernova that accompanies a gamma-ray burst, where the stellar core collapses into a black hole, highly energetic jets of plasma, made up of electromagnetic radiation and excited subatomic particles, are emitted at the poles of the star's remnants. These jets travel at close to the speed of light and planet Earth appears to be in the path of this plasma jet. Evidence from the image taken of the star shows that one of WR104's poles points in our direction, with an inclination of around 12 degrees from the front line of sight. However, due to the difficulty of obtaining precise measurements, the degree of inclination of the star's pole can vary between 0 and 16 degrees.

If planet Earth were to be hit by this highly energetic plasma jet, the planet would not be destroyed, but there would be a mass extinction of the living beings that inhabit the planet. Because the electromagnetic radiation coming from the stellar jet would drastically alter the atmosphere, where 30 per cent of the ozone layer would be destroyed, allowing a high incidence of ultraviolet rays from the sun, causing alterations in the DNA of various living beings that could suffer mutations incompatible with life, such as cancer in human beings.

In addition, smog (dark gas) could form as a result of the degradation of ozone (O_3) together with molecular nitrogen (N_2), resulting in nitrogen dioxide (NO_2). Smog has a brown colour, which would make it difficult for plants to capture the sunlight they need for photosynthesis, and it could also precipitate with the water in the atmosphere, causing acid rain, which would cause corrosion in both natural and man-made materials.

And, in addition to the problems arising from the electromagnetic radiation of the plasma jet, the highly excited subatomic particles present in the jet would cause more drastic effects, as they are highly radioactive, and when they hit a certain region of the planet, it would be practically sterilised.

The WR 104 star can therefore be considered a giant star that could become apocalyptic when it goes supernova. Although this is possible, there are still no concrete reasons to worry, since there are many uncertainties about this star, such as: the degree of inclination of its pole with Earth, whether its mass is sufficient to collapse into a black hole and suffer a gamma-ray burst, and its real distance from planet Earth, estimated at 8,000 light-years, but which could be closer or further away. For all these reasons, the WR 104 star is, for the time being, just one object in the universe with many mysteries to be revealed.

CHAPTER 6

FINAL CONSIDERATIONS

In the Main Sequence, the star, regardless of its initial mass, fuses H into He until the H runs out and the star starts to produce new elements, such as C, N and O, until it obtains Fe in its core.

After this phase, low-mass stars like the Sun become Red Giants, gradually and slowly losing mass into space until a hot, dense core remains, known as a white dwarf.

More massive stars such as Wolf-Rayet have a much faster evolution and undergo a large supernova explosion (SN), which promotes the formation of even heavier elements, and often releases jets of gamma radiation from their poles of rotation, which could be harmful to any celestial object in its path.

When comparing the Sun with WR104, it can be concluded that the object that poses the greatest risk to planet Earth is the giant star WR 104 because, as its data is uncertain, it is not known exactly how old it is or what stage of evolution it is in.

If it has already passed into its final stages, we can expect a very large amount of gamma radiation to reach us at any moment, which could decimate life on our planet.

The Sun, however, will reach its Red Giant stage in approximately 5 billion years.

At that point, the Earth will be scorched by its expansion and the planet will become extinct. Fortunately, there will be plenty of time for the human race to seek solutions off Earth, if we survive as a species until then.

BIBLIOGRAPHICAL REFERENCES

BERGMAN, J. Sunspots. 2016. Available at: http://www.windows2universe.org/sun/atmosphere/sunspots.html . Accessed on: 17 June 2016.

BERGMANN, TS et al. **Black holes**. 2011. Available at: http://www.if.ufrgs.br/~thaisa/bn/bn main.htm . Accessed on 27 May 2016.

BURROWS, S. A.; LATTIMER, J. M. **The birth of neutron star**. The Astrophysical Journal. v. 307, pp. 178-196, 1986.

CAIBO, S. R. **20th century -** Astronomy and astronautics - stellar evolution. Available at: http://www.cdcc. usp. br/cda/sessao-astronomia/seculoxx/textos/a-evolucao-estelar.htm. Accessed on: 28 Dec. 2011.

CASAS, R. **Black hole**. 1999. Available at: http://www.observatorio.ufmg.br/pas19.htm . Accessed on 02 June 2016.

CASTRO, Ricardo de. **How does the alchemy of the stars work?** The nucleosynthesis of chemical elements in the Universe. 2010. Available at: http://eternosaprendizes.com/2010/08/17/como-funciona-a-alquimia-das- stars-the-nucleosynthesis-of-chemical-elements-in-the-universe/ . Accessed on: 28 February 2012.

CHAVES, A. **Stellar evolution. 2011.** Available at: http://www.observatorio.ufmg.br/Pas104.htm. Accessed on: 04 September 2012.

CHAVES, A. **Supernova the catastrophic death of large stars.** 2011. Available at: http://www.observatorio.ufmg.br/Pas105.htm . Accessed on: 28 May 2016.

CLAYTON, D. D. **Principles of stellar evolution and nucleosynthesis**. New

York: McGraw-Hill Book Company, 1968.

FARIAS, T. M. B. **Study of the thermoluminescent and optical absorption properties of eight varieties of quartz**. São Paulo, 2008. 81f. Thesis (Doctorate) - University of São Paulo. Institute of Physics. Available at: http://www.teses.usp.br/teses/disponiveis/43/43134/tde-15042009-165630/en-br.php . Accessed on: 13 January 2012.

FELIX, Regina Celia Medeiros. **Study to determine the parameters describing the dynamics of a galactic supernova using a future neutrino detector**. 2009. Dissertation (Master's in Physics) - Institute of Physics, University of São Paulo, São Paulo, 2009. Available at: http://www.teses.usp.br/teses/disponiveis/43/43134/tde-17082009-084349/ . Accessed on: 28 May 2016.

FILGUEIRAS, C. Spectroscopy and chemistry: the discovery of new elements at the threshold of quantum theory. **Química Nova,** p. 22-25. n. 3, 1996.

FILHO, KSO; SARAIVA, MFO. **Final evolution of the stars**. 2016. Available at: http://astro.if.ufrgs.br/estrelas/node14.htm . Accessed on: 31 May 2016.

FONSECA, A. M. M. **Electro-cyclotronic emission in the TCABR tokamak:** an experimental study. São Paulo, 2005. 211f. Thesis (Doctorate) - University of São Paulo. Institute of Physics - Department of Applied Physics. Available at: http://www.teses.usp.br/teses/disponiveis/43/43134/tde-29092005-132047/pt-br.php. Accessed on: 13 January 2012.

FRIAÇA, A. C.S; DAL PINO, E.; SODRÉ Jr., L.; JATENCO-PEREIRA, V. **Astronomy: an overview of the universe**. 2.ed. São Paulo: Editora da Universidade de São Paulo, 2008.

GONCALVES, Denise R. **Not every star is young**. Cienc. Cult., São Paulo, v. 61, n. 4, 2009. Available at: http://cienciaecultura.bvs.br/scielo.php?script=sci arttext&pid=S0009-67252009000400012&lng=en&nrm=iso . Accessed on: 28 May 2016.

GREGORIO-HETEM, J. ; JATENCO-PEREIRA, V. **Fundamentals of astronomy**. Available at: http://www.telescopiosnaescola.pro.br/aga215/aga215.html. Accessed on: 08 September 2012.

JACOBSEN, R. B**. Plasma of Quarks and Gluons inside Neutron Stars**. Dissertation (Master's in Physics; Federal University of Rio Grande do Sul, Porto Alegre, 2007).

KAUFMANN III, W. J; Comins, N,.F,. **Discovering the Universe**. 8.ed. Porto Alegre: Bookman, 2010.

KAUFMANN III, W. J; FREEDMAN, R. A. **Universe**. 8.ed. New York: W. H. Freeman and Company, 2008.

LARANJEIRAS, C.C. **Spectroscopy: investigating the intimate structure of matter**. 2010. Available at: http://www.experimentum.org/blog/?p=567 . Accessed on: 20 September 2012.

MAZUREK, T. J. ; WHEELER, J. C. **Fundamentals of cosmic physics**. New York: V. Canudo, 1980.

NASA. **Supernova remnant W49B**. 2013. Available at: http://www.nasa.gov/mission pages/chandra/multimedia/w49b.html . Accessed on 25 May 2016.

NASA. Astronomy Picture of the Day: WR104; 2014. Available at: https://apod.nasa.gov/apod/ap140603.html . Accessed on: 09 April 2017.

NASA. **Crab Pulsar Wind Nebula.** 2008. Available at: http://apod. nasa.gov/apod/ap081227.htm l. Accessed on: 6 June 2016.

NASA. **Layers of the sun**. 2007. Available at: http://www.nasa.gov/mission pages/hinode/solar 020.html Accessed on: 17 June 2016.

NASA. **Supernova Remnant and Neutron Star**. 1998. Available at: http://apod.nasa.gov/apod/ap980425.html . Accessed on: 06 June 2016.

NERLICH, S. Universe today: astronomy without a telescope - archemy by supernova. 2015. Available at: http://www.universetoday.com/70646/astronomy-without-a-telescope-alchemy-by-supernova/. Accessed on: 18 January 2016.

NEYER, R. Neutron Star, 2007. Available at:< http://www.nasa.gov/mission pages/GLAST/science/neutron stars.html> Accessed on: 6 June 2016.

NOEL, D. Inside the Earth - the heartfire model. 2012. Available at:http://www.aoi.com.au/bcw/Heartfire/. Accessed on: 04 June 2016.

THE UNIVERSE: Secrets of the Sun, Ep.1. Directed by Douglas J. Cohen. Production: **The History Channel**. Distribution: LOG ON Editora Multimídia. 2007.

OLIVEIRA FILHO, K. S.; SARAIVA, M.F.O. **Astronomy and Astrophysics**. 2016.
Available at: http://astro.if.ufrgs.br/index.htm. Accessed on: 19 July 2016.

OLIVEIRA FILHO, K. S.; SARAIVA, M.F.O.**Astronomia e Astrofísica**. 2. ed. São Paulo: Livraria da Física, 2004.

PENATE, L et al. **Short-term effects of gamma ray bursts on oceanic photosynthesis**. Astrophysics and Space Science. V.330, n.2, pp.211-217,

2010.

PHILLIPS, T. **Discovered: A New Kind of Pulsar**. 2008. Available at: http://science.nasa.gov/science-news/science-at-nasa/2008/17oct gammaraypulsar/. Accessed on: 6 June 2016.

PLAIT, P. **WR 104: A nearby gamma-ray burst?** In: Discover magazine. New York: Corey S. Powell, 2008. Available at: http://blogs.discovermagazine.com/badastronomy/2008/03/03/wr-104-a-nearby-gamma-ray-burst/. Accessed on: 17 July 2012.

RAMOS, P.S. (transl.) **Fusion in the Universe:** the origin of jewellery, 2007. Available at:http://www.scienceinschool.org/2007/issue5/fusion/portuguese. Accessed on: 21 May 2012.

REES, M. **O Reino das Galáxias**. São Paulo: Duetto Editorial, 2008.

REIS, N.T.O.; ANDREOLLA, C.V.; SOUZA, M.O.;SANTOS, N.S. **Space Weather.** In: Astronomy, Astronautics and Space Sciences at School. Available at:http://educacaoespacial.wordpress.com/recursos-2/conteudos/formacao-2/clima-espacial/. Available on: 20 September 2012.

SANTO, M. G. **Study of nuclear reactions of astrophysical interest using the trojan horse method**. São Paulo, 2009. 180f. Thesis (Doctorate) - University of São Paulo. Institute of Physics. Available at: http://www.teses.usp.br/teses/disponiveis/43/43134/tde-07042009-161106/pt-br.php. Accessed on: 03 March 2012.

SANTOS, T. J. **Formation of the chemical elements:** from the great explosion to the stars. Available at: http://www.observatorio.ufmg.br/pas36.html. Accessed on: 28 December 2011.

SANTOS, T. J. **Sun**: structure and evolution. In: UFMG Frei Rosário Astronomical Observatory. Available at: http://www.observatorio.ufmg.br/pas25.html Accessed on: 04 September 2012.

SERVULO, F. **Special 100 years of the theory of general relativity**. 2015. Available at: http://www.misteriosdouniverso.net/2015/03/especial-de-100-anos-da-teoria-da.html. Accessed on 15 May 2016.

SÉRVULO, F. Astronomical events - supernovae. **Mysteries of the Universe**. 2014. Available at: http://www.misteriosdouniverso.net/2014/05/eventos-astronomicos-as-supernovas.html. Accessed on: 04 June 2016.

SILVA, A. V. R. **Our star**: the sun. 1. ed. São Paulo: Livraria da Física, 2006.

TAURINES, A. R. **Neutron stars in relativistic mean field theories**. Dissertation (Master's Degree in Physics; Federal University of Rio Grande do Sul, Porto Alegre, 1999).

TUTHILL, P. et al. The Prototype Colliding-Wind Pinwheel WR 104. **The Astrophysical Journal**. p. 698-710, 2008.

TUTHILL, P. **The twisted tale of wolf-rayet 104 first of the pinwheel nebulaebula**. In: The University of Sydney. Sydney: School of Physics, 2009. Available at: http://www.physics.usyd.edu.au/~gekko/wr104.html. Accessed on: 25 July 2012.

TUTHILL, P. **WR 104: The prototype pinwheel nebula**. In: The University of Sydney. Sydney: School of Physics, 2009. Available at: http://www.physics.usyd.edu.au/~gekko/pinwheel.html. Accessed on: 20 July 2012.

Printed by Books on Demand GmbH, Norderstedt / Germany